Cambridge Elements

Elements of Paleontology
edited by
Colin D. Sumrall
*University of Tennessee*

# ECHINODERM MORPHOLOGICAL DISPARITY: METHODS, PATTERNS, AND POSSIBILITIES

Bradley Deline
*University of West Georgia*

**CAMBRIDGE**
**UNIVERSITY PRESS**

University Printing House, Cambridge CB2 8BS, United Kingdom

One Liberty Plaza, 20th Floor, New York, NY 10006, USA

477 Williamstown Road, Port Melbourne, VIC 3207, Australia

314–321, 3rd Floor, Plot 3, Splendor Forum, Jasola District Centre, New Delhi – 110025, India

79 Anson Road, #06–04/06, Singapore 079906

Cambridge University Press is part of the University of Cambridge.

It furthers the University's mission by disseminating knowledge in the pursuit of education, learning, and research at the highest international levels of excellence.

www.cambridge.org
Information on this title: www.cambridge.org/9781108794749
DOI: 10.1017/9781108881883



First published 2021

*A catalogue record for this publication is available from the British Library.*

ISBN 978-1-108-79474-9 Paperback
ISSN 2517-780X (online)
ISSN 2517-7796 (print)

# Echinoderm Morphological Disparity: Methods, Patterns, and Possibilities

Elements of Paleontology

DOI: 10.1017/9781108881883
First published online: January 2021

---

Bradley Deline
*University of West Georgia*

**Author for correspondence:** Bradley Deline, bdeline@westga.edu

**Abstract:** The quantification of morphology through time is a vital tool in elucidating macroevolutionary patterns. Studies of disparity require intense effort but can provide insights beyond those gained using other methodologies. Over the last several decades, studies of disparity have proliferated, often using echinoderms as a model organism. Echinoderms have been used to study the methodology of disparity analyses and potential biases as well as documenting the morphological patterns observed in clades through time. Combining morphological studies with phylogenetic analyses or other disparate data sets allows for the testing of detailed and far-reaching evolutionary hypotheses.

**Keywords:** *Echinodermata*, disparity, morphology, morphospace, phylomorphospace



ISBNs: 9781108794749 (PB), 9781108881883 (OC)
ISSNs: 2517-780X (online), 2517-7796 (print)

# Contents

# 1 Introduction

One of the main goals within paleontology is the elucidation of large-scale evolutionary patterns as well as gaining insights into the underlying mechanisms. There are three primary methodologies utilized to explore evolutionary dynamics through time: reconstructing phylogenetic relationships, compiling taxonomic biodiversity, and quantifying morphologic diversity. The first two metrics largely focus on one particular aspect of evolution, speciation. Phylogenetics attempts to reconstruct the splitting of taxonomic groups into discrete units. Although the data collected for phylogenetic reconstructions can be co-opted for other purposes, the primary aim is identifying the relationships between taxa. Biodiversity is driven by the rate of origination of new taxa as well as the rate of extinction. Obviously, biodiversity is a complex metric that is biased by a number of different factors but it is largely capturing the splitting and truncation of evolutionary lineages. The quantification of morphological diversity (i.e. disparity), captures many aspects of evolutionary change including speciation, the degree of change during speciation, and the amount of anatomical change within a lineage. In addition, the metric captures extinction events as well as the properties of extinction, such as selectivity towards particular morphological forms or features. Given the amount of information contained within this metric, studies examining disparity are work intensive and time consuming, there are many potential biases, and the resulting patterns are often difficult to interpret. However, this methodology provides a broad and encompassing view of evolutionary change within and between lineages through time.

The quantification of organismal form has long been recognized as a valuable tool to explore evolutionary dynamics (Thompson, 1917; 1942). However, the mathematical exploration of anatomy only became commonplace with the rise of computational ability given the inherent mathematical complexity of organismal form (Raup, 1962). These computational methods of quantifying organismal form were first applied to fossil organisms in a series of papers describing the coiling properties in mollusks including snails and ammonoids (Raup, 1962; Raup and Michelson, 1965; Raup, 1966; Raup, 1967). Raup and Michelson (1965) quantified the geometry of shell coiling based on four distinctive properties and constructed a multidimensional space of theoretical forms. This morphospace of coiling shells could then be described in terms of both theoretically plausible forms as well as those observed in nature, allowing the formulation of hypotheses on the factors limiting the realized forms within morphospace. The areas occupied by different taxonomic groups were plotted to examine the different constraining factors and shared evolutionary trajectories between related taxa (Raup, 1967).

In the following decades, paleontologists increasingly stressed the importance of detailed examinations of the diversity of organismal form in addition to biodiversity. Initially, studies attempted to explore disparity by contrasting the biodiversity at multiple taxonomic ranks, that is, using class or phylum diversity as a proxy for disparity. The use of taxonomic rank as a proxy for disparity was applied broadly by Valentine (1969), exploring the taxonomic and ecological structure of the marine benthos through the Phanerozoic. Valentine (1969) found an inverse relationship between higher- and lower-order taxonomic diversity in marine organisms, which he attributed to an increased ecological specialization. This specialization canalized morphology and prevented large departures and morphological innovations later in the Phanerozoic. Yochelson (1978; 1979) used class-level diversity as a proxy for disparity, interpreting the origin of new classes as the appearance of major anatomical changes. These morphological shifts thus enabled adaptive radiations into new ecological habits. Following this, Jaanusson (1981) suggested that morphological innovation occurred stepwise, with each advancement crossing functional thresholds enabling progressive diversifications. Jaanusson (1981) provided multiple examples, such as shifts in brachiopod detention, which qualitatively suggested a disjunction between taxonomic and morphological diversity and highlighted the importance of disparity in studies of macroevolution. Runnegar (1987) further defined disparity as 'the amount of difference between related phyla, classes, species, individuals, proteins, genes etc' (p. 41). Furthermore, Runnegar (1987) suggested that disparity was best explored qualitatively to capture the innovative changes, such as those discussed by Jannusson (1981).

Alternatively, many studies attempted to apply quantitative approaches to estimations of disparity. Overall, these methods lagged behind taxonomic-based approaches for many reasons. There are many features within an organism that could potentially be quantified and several different metrics to attempt to capture the range of forms being characterized. For instance, Cherry et al. (1982) explored disparity within 184 vertebrate taxa based on a relatively small set of linear measurements. They found an equitable degree of morphological variation within genera of amphibians, lizards, and mammals. However, the equivalence of taxonomic level and morphological variation broke down at higher taxonomic ranks, which questions the direct use of class or phylum diversity as a meaningful proxy of morphologic diversity (Cherry et al., 1982). Therefore, the need for quantifying morphology independent of relying on diversity of higher taxonomic biodiversity became apparent. This quantification of body plans was popularized with the re-description of the Burgess Shale fauna and its importance in understanding the evolution of animal body plans. Gould (1989; 1991) wrote extensively regarding the hypothesis that disparity followed a very different pattern from diversity through the Phanerozoic and proposed an early peak (Cambrian) in diversity of body plans. Gould (1991, p. 441) stated 'the claim for greater early disparity cannot be confidently established until we develop quantitative techniques for the characterization of morphospace and its differential filling through time'. This work spurned the large-scale collection of morphological data to test these proposed patterns of macroevolution.

Given their diversity and unique morphologies within the Burgess Shale, arthropods were often used to test the idea of an early peak in morphological disparity. Briggs et al. (1992) constructed a data matrix of 134 characteristics to analyze Cambrian and modern arthropods and found equitable levels of disparity. Based on this, they concluded that the view of Cambrian disparity was clouded by an artefact of taxonomy and the pull of the unusual. However, they also noted that the filling of the morphospace occurred rapidly, such that the Cambrian Explosion was dampened but still a pivotal event in the history of life (Briggs et al., 1992). This work was followed by a similar study on priapulids, in which Wills (1998) found a decrease in disparity between the morphologically distinctive (i.e. non-overlapping in morphospace) Cambrian and modern worms. Again, this both strengthened the hypothesis of an initial burst of morphological innovation in the Cambrian and challenged the decimation of disparity through time (Wills, 1998).

The focus of disparity studies on the Burgess Shale fauna following the publication of Gould's *Wonderful Life* (1989) de-emphasized the importance of echinoderms because of their relative scarcity within the Burgess Shale fauna (Sprinkle and Collins, 2006; Zhao et al., 2010). However, echinoderms soon became model organisms for the study of morphological disparity. Echinoderms are ideal subjects for studies in disparity in that they were and are ecologically and taxonomically diverse, in addition to the expansive range of morphological features and body plans within the phylum (Paul and Smith, 1984). Importantly, most echinoderms are highly skeletonized, which produces fossils that are character-rich with potentially relatively little external morphology lost from taphonomic processes compared to other phyla (Brett et al., 1997; Deline and Thomka, 2017). Finally, echinoderms have an incredible richness of higher-order taxonomic groups in the early Paleozoic compared to today (Sumrall and Wray, 2007). This early burst of body plans places echinoderms as a group most likely to conform to the initial peak in disparity hypothesis.

Early studies of echinoderm morphological disparity confirmed the diversity and complexity of the phylum. Foote (1991) compiled a geometric morphometric data set of Paleozoic blastoids and found a significant disconnect between taxonomic and morphologic diversity. Blastoids steadily increased in disparity from the Silurian through the Permian, highlighting their continued morphological plasticity and capacity for innovation. Overall, this provided a counterargument to the patterns seen in arthropods and priapulids with Permian blastoids occupying a broader range of morphospace right before the extinction of the clade (Foote, 1991). Foote (1992) then explored blastozoan evolution by constructing a discrete character matrix. This study showed that the ratio of disparity to diversity was highest in the Cambrian, but disparity grew steadily through the early Paleozoic peaking in the late Ordovician (Foote, 1992). This corroborated the hypothesis of an initial explosion of disparity in the Cambrian paired with continued evolutionary innovation within the clade. These early studies showed the potential for morphological studies within echinoderms as well as the capacity for continued morphological exploration and innovation within echinoderms. These studies also stressed the need to examine disparity beyond the Cambrian in many different echinoderm groups, using different methodologies, and across variable taxonomic scales (Foote, 1997).

Over the past 30 years, studies of echinoderm disparity have stretched across the phylum to explore different methods for estimating trends in disparity, potential biases in the quantification of morphology, and how taphonomy can alter these perceived trends. Studies of echinoderm disparity have highlighted the rapid initial morphological diversification within clades but also the importance of continued constraint through time, particularly following mass extinctions or during faunal turnover events. In addition, multiple studies have explored the underlying developmental and biologic factors enabling and constraining morphology through time. Deline et al. (2018) placed echinoderms within a broader framework of metazoan morphology. This study used the work of Ax (1996; 2000; 2003) to construct a morphospace of extant metazoans, which was then expanded to include a snapshot of the diversity of Cambrian animals. Overall, echinoderms cover a small area within metazoan disparity (Figure 1A). Even though sampling was correlated with genus-level taxonomic diversity, echinoderms were relatively undersampled, thus reducing their apparent morphological importance. This was further emphasized by focusing on the modern (six clades) and Cambrian (nine genera), neither of which highlight the peak in echinoderm body-plan diversity in the Late Ordovician (e.g. peak in class-level diversity). Furthermore, Deline et al. (2020) compiled an expansive data set of early Paleozoic echinoderm morphology recovering four major body plans during the initial explosion of echinoderm morphology (Figure 1B). In addition, many studies have explored morphological patterns within echinoderms at the class or subclass level (e.g. Lefebvre et al., 2006; Deline et al.,

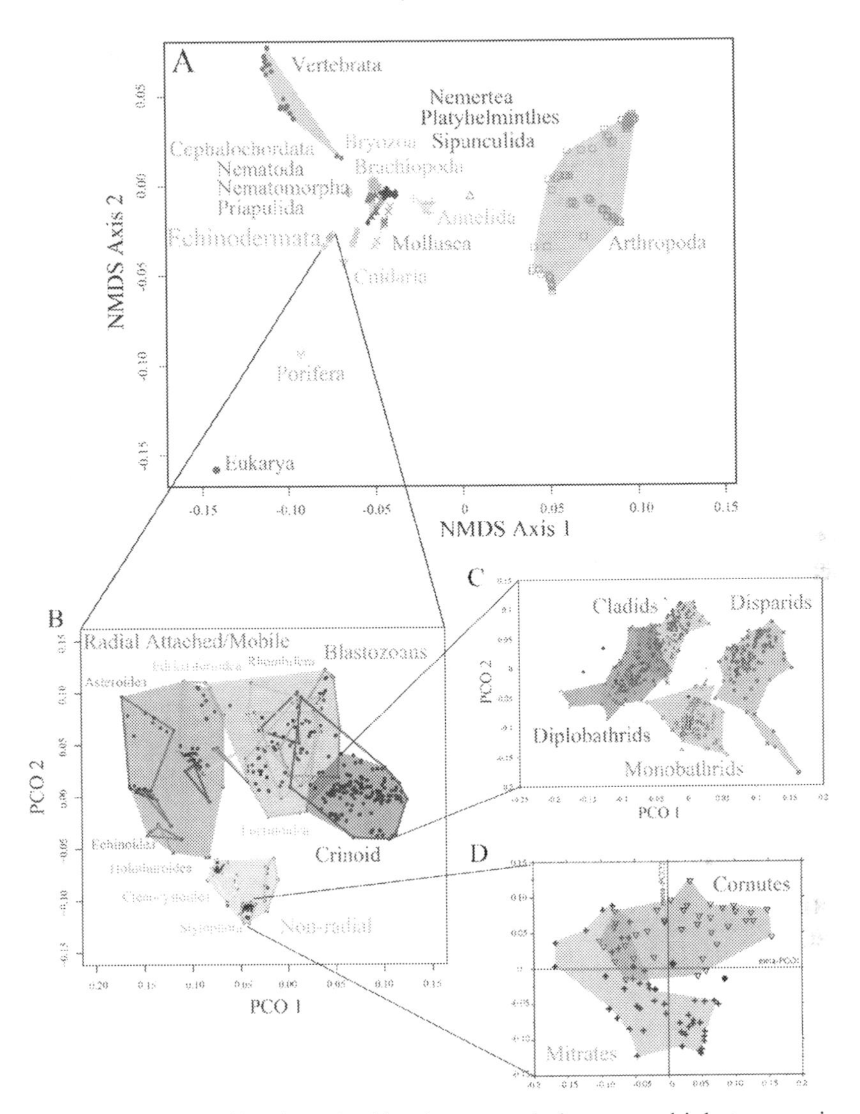


**Figure 1** The quantification of echinoderm morphology at multiple taxonomic scales from kingdom (metazoans) to phylum to class (crinoids and stylophorans). Morphospaces modified from Lefebvre et al. (2006), Deline et al. (2012), Deline et al. (2018), and Deline et al. (2020).

2012; Figure 1C, D) documenting detailed patterns of morphological change through time. These studies across different taxonomic levels allow an expansive prospective of echinoderm morphological evolution within and beyond the phylum.

Foote (1997) reviewed the progress of studies of morphological disparity and highlighted future directions. Studies of echinoderm disparity have helped to fill some of the prominent gaps in our knowledge. However, there are still ample opportunities to utilize morphology to better understand the evolution of echinoderms. The current review aims to explore the diverse methods that have been utilized to quantify morphological evolution in echinoderms as well as the choices and biases that alter perceived trends in disparity through time. Given the diversity of methods and experimental designs, several prominent patterns have been repeatedly found which will be discussed along with their important developmental and macroevolutionary implications. Finally, promising directions of study utilizing echinoderm morphology will be highlighted.

## 2 Methods of Quantifying Morphology

The quantification of morphology and estimates of disparity can be accomplished using multiple techniques (Hopkins and Gerber, 2017). Broadly, these methods can be broken down into two major groups: morphometric approaches utilizing continuous measures and categorical approaches utilizing discrete characters.

### Morphometric Methods

Morphometric approaches have the benefit of being more intuitive in that the metrics directly correspond to easily visualized shape parameters. Common morphometric methods include the direct measurement of features (i.e. traditional morphometrics), comparisons of the positions of distinctive landmarks (i.e. geometric morphometrics), or the quantification of the overall outline of features or the entire organism (Webster and Sheets, 2010). The choice of the morphometric methodology depends on the organism, taxonomic scale of study, and hypothesis being addressed. Obviously, the level of anatomical detail that can be captured decreases with an increase in taxonomic scope of study. In addition, certain body plans lend themselves to specific methodologies. For example, the overall body of flattened taxa such as stylophorans or cinctans could be successfully characterized with outline or landmark analysis. Overall, all of these methods have been utilized in the study of echinoderm disparity.

Lefebvre et al. (2006) explored the morphological diversity within stylophorans using a traditional morphometric approach of directly measuring features. This group contains a wide array of characteristics and large variability in overall body plans. Therefore, this is an enticing group to explore morphologically, but the puzzling anatomy makes quantification difficult. The low profile of the theca allowed a two-dimensional quantification of the overall body shape as well as the geometry of individual plates. The direct comparison of individual thecal elements requires a study at lower taxonomic rank and careful analysis of potential plate homologies across the group. To accurately quantify shape, they applied measurements such as circularity of the theca and individual plates as well as the relative area of individual plates or features compared with the area of the entire theca. Using ratios of features effectively removed body size from the analysis to obtain a clearer measure of the overall form. All told, this produced over 70 variables for both the upper and lower surface of the theca that accurately quantified body shape as well as the constituent pieces, which would not have been possible with other methods such as Procrustes-based landmark analysis (Lefebvre et al., 2006).

Landmark-based geometric morphometrics has been utilized many times to differentiate species, assess ontogenetic change, or document changes in disparity through time. As with all of the methodologies discussed, the a priori choices are pivotal to the results of the study. Ideally, each individual landmark should represent an easy-to-recognize homologous point (Figure 2). Within echinoderms, the clearest landmarks would appear at plate junctions, such that the placement of the landmark is accurate and reproducible. This level of anatomical similarity across samples again requires a lower taxonomic breadth of study, but this methodology allows for the easy visualization of the forms

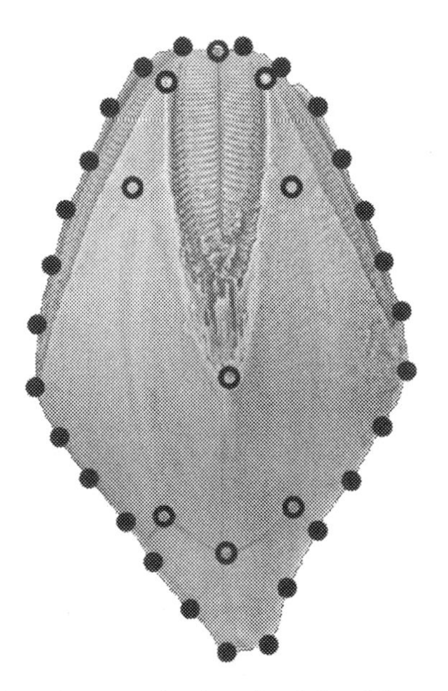

**Figure 2** A comparison between landmarks defined by specific plate junctions (open circles) as used by Foote (1992) and semi-landmarks used to define a thecal outline (closed circles) as used by MacLeod (2015) shown on the blastoid *Pentremites meganae* (Atwood and Sumrall 2012).

being studied. Foote (1991) used this approach to study disparity in blastoids through the Paleozoic. This methodology is well-suited for this clade in that the reduction in thecal plates makes the recognition of plate homology easier. Landmarks were selected across the theca at plate junctions from three different perspectives: basal, oral surface, and theca profile. These landmarks were standardized to eliminate differences in size to exclusively compare the three-dimensional shape of the theca. Eble (2000) used similar methods to explore differences in disparity between sister clades of atelostomate echinoids. In this case, landmarks were chosen across the apical system, ambulacra, body outline, as well as the peristome and periproct. Similar to the blastoid study, this provided a three-dimensional geometric prospective of the echinoid test independent of size. In addition, the use of mostly non-plate junction landmarks allows for a broader taxonomic comparison. Importantly, Eble (2000) notes that while this methodology captures significant aspects of echinoid morphology, it is not exhaustive. Aspects of the spine turbercles, overall plating, and ambulacra pore structures were not characterized. Obviously, all of these methods quantify morphology based on a subset of features, but it is worthwhile to consider whether the patterns being characterized are representative of the overall morphology or exclusively the features being examined. Studies in other taxonomic groups (i.e. trilobites) have indicated that morphology of one body region can be used as a proxy for overall morphology (Hopkins, 2017), but that is likely far from universal and needs to be further tested in other taxonomic groups (see Deline and Ausich, 2017).

In many echinoderm groups, the body plating is extraordinarily variable such that the recognition of homologous points might be limited to the body openings, which would be morphologically uninformative. In these cases, methods characterizing the body shape or outline might be more applicable than the identification of distinctive landmarks. Characterizing the shape of a body or feature can be accomplished through tracing the feature, defining the shape through Fourier analysis, or by placing semi-landmarks (Figure 2) around the outline that can be compared between specimens (Webster and Sheets, 2010). MacLeod (2015) used a combined landmark and outline analysis method to characterize morphology within the eocrinoid *Gogia*. Gogiid eocrinoids are difficult to characterize morphologically because of a lack of distinctive and homologous thecal plates and a propensity for severe taphonomic alteration (e.g. crushing). MacLeod (2015) characterized gogiid morphology using four ‘distinctive landmarks’ that defined the aboral cup as well as a series of semi-landmarks that defined the outline of the lower theca. Although there are significant issues documenting standard orientations in gogiids that lack perfect radial symmetry, this method could potentially capture the shift in body outline from the stalk to the theca in a broad sense. Using these landmarks separately and in combination, MacLeod (2015) suggested he was able to extract biologically meaningful information from the characterized individuals such as ontogenetic shifts in thecal morphology. Although more research is warranted, this study highlights that morphometrics can potentially be applied broadly, even within groups that would initially appear to be poor candidates for this type of analysis. However, as with other morphometric techniques, the taxonomic breadth can be fairly limited in that thecal or other outlines constructed with different underlying plates would not be a meaningful comparison.

In addition to these primary methods, there are other methodologies that could yield promising results in the characterization of morphology and assessing patterns of disparity through time. Echinoderm skeletons that grow through a combination of plate growth and addition allow mathematical modelling of growth and development (Zachos and Sprinkle, 2011). Detailed characterization of different growth models allows for the potential quantification of developmental morphology assessing a different aspect of body form. A similar approach was utilized by Hoyal Cuthill and Hunter (2020), that quantified crinoid calyx morphology based on graph theory. This method enables the compilation of theoretical forms and their structural implications, thus providing a way of assessing the functional implications of complex plating patterns (Hoyal Cuthill and Hunter, 2020). Geometric form can also be characterized in much greater detail with the construction of three-dimensional digital models with GIS-based methodologies (e.g. Sheffield et al., 2012) or x-ray tomography (e.g. Zamora et al., 2012). However, the increased level of morphological detail in these methods often comes with a greater financial cost (e.g. specialized software, increased computation requirements, or equipment) as well as requiring significantly more time and effort, which would severely limit the scope of studies utilizing these techniques.

## Character-Based Methods

If the aim of the study is broader than can be achieved through the use of morphometric methods, or if taphonomic alteration makes those methods inappropriate, morphology can then be quantified through the use of discrete characters. Ideally, characters are chosen to cover all aspects of the organism's morphology, including convergent traits and autapomorphies. Many recent studies have utilized cladistic data sets for this purpose, but most character suites constructed for that purpose focus on phylogenetically important traits and intentionally avoid the confounding influence of homoplastic traits or phylogenetically uninformative autapomorphies. The exclusion of these aspects of overall morphology would then create similar issues to those discussed earlier in regard to landmarks, potentially skewing the morphological patterns away from whole-organism characterization. Although, comparing character states captures aspects of morphology, especially at broader taxonomic scales, the meaning of the distances between organisms becomes murkier. With morphometrics, the resulting differences are easy to visualize and potentially equitable in the aspects of morphology they capture, but each discrete character is unique in terms of morphology being described. In addition, the genetic or developmental processes leading to the changes in geometric shape are potentially more direct than with the appearance of a novel trait, a change in character state, or shifts in characters broadly describing shape. Therefore, the distances between taxa using discrete characters should be seen as amalgamations of vastly different types of characters with unclear relationships to the underlying developmental mechanisms. That in no way invalidates the use of discrete characters but it necessitates caution in the interpretation of the constructed morphospaces in that the meaning of distances are then more abstract and likely to have affine rather than metric properties (Mitteroecker and Huttegger, 2009; Huttegger and Mitteroecker, 2011).

Foote (1992) constructed a character suite describing morphology within blastozoan echinoderms. Characters were chosen without regard to their known or presumed phylogenetic importance. However, he notes that there is often significant overlap between characters used for the purposes of capturing disparity and deciphering phylogenetic relationships. In total, Foote (1992) used 65 binary and multistate characters that described features across the entire body from attachment structures to characteristics of the brachioles. In coding the characters, Foote (1992) focused on topological position rather than homology to attempt to avoid an overt phylogenetic signal. Obviously, this choice has ramifications in the resulting patterns of disparity. If features like stalks evolve multiple times and were thus coded as independent features, that would increase the estimation of disparity, even though the two stalks are similar topologically and functionally such that they should occupy similar areas of morphospace. This again highlights the importance of caution in the straightforward use of cladistic data sets for the purpose of describing disparity.

Following this work, Foote (1994a; 1994b; 1995a; 1995b; 1999), in a series of papers, explored the morphology of crinoids throughout the Phanerozoic. Even though this is a lower taxonomic group, the high biodiversity, long geologic history, and diversity of forms within crinoids presented unique challenges. The wide array of forms within crinoids presents the dichotomy of being an ideal group to study patterns of disparity while at the same time creating difficulties constructing a meaningful estimation of morphologic differences between taxa. Essentially, there needs to be a coded and recognizable commonality between the taxa to aid in the construction of the morphospace. In other words, if all of the features in one organism are non-applicable in another, the comparison between the two is essentially meaningless. In the construction of the character suite, Foote (1994a; 1994b; 1995a; 1995b; 1999) included characters that described the structure of the aboral cup that are inherently focusing on homology between the clades. He states, 'The present scheme of character coding thus speaks out of both sides of its mouth, emphasizing homology in some cases and convergence in others' (Foote 1999, p. 3). Overall, this needed compromise allowed the comparison of dramatically different organisms in a meaningful way while attempting to dampen the overt phylogenetic signal. To explore how the choice of one of many homology schemes influenced patterns of disparity, Foote (1999) recoded taxa using other schemes and found this a priori choice had little effect on the ensuing results.

One potential issue that can arise is recognizing the differences between traits that are unknown because of lack of preservation compared with traits that are non-applicable for a particular animal. For instance, a character describing the type of branching in crinoid arms would not be applicable for a crinoid with unbranching arms. In this case, coding the trait as missing (arms are unpreserved) or non-applicable (arms are unbranched) have very different meanings and lumping these character states together would significantly increase the amount of missing data. Deline (2009) addressed this issue by coding unpreserved data as missing and coding non-applicable data as zero with applicable character states as progressive integers. Using this coding scheme along with Gower's similarity coefficient allowed treating these character states as distinctive. The end effect of this applied to an independently coded version of Foote's character suite was increasing the distance between the major groups of crinoids. Essentially, using this scheme incorporated mismatched characters between taxa with applicable and non-applicable states, which then incorporated a hierarchical structure to the data set.

Using a similar philosophy as Foote (1999), Deline (2015) and Deline et al. (2020) constructed a character suite that encompassed the range of morphology found within early Paleozoic echinoderms. To meaningfully compare disparate groups of echinoderms, there need to be some characters that are broadly applicable. To do this, characters were chosen that were expected to be broadly convergent (e.g. body orientation relative to the sediment or mobility) as well as those that likely contain a strong phylogenetic signal based on proposed hypotheses of homology (e.g. Mooi and David, 1997; Sumrall and Waters, 2012; Kammer et al., 2013). Again, this compromise gave structure to the resulting morphospace (Figure 1B) without overtly focusing on the phylogenetic signal. The data set was also constructed in a hierarchical manner following Deline (2009), which allowed the direct testing of hypotheses regarding the relative plasticity of characters during the initial morphological diversification of echinoderms (Deline et. al., 2020). These types of analyses allow the examinations of much broader hypotheses but also can have potential issues as will be discussed subsequently.

Once the character-based morphospace is constructed, there are multiple metrics that have been proposed to explore disparity patterns through time. Ciampaglio et al. (2001) compared seven different metrics in terms of the perceived patterns of morphological evolution as well as their sensitively to small sample sizes and missing data. They found that the metrics differed in their sensitivities to shifts in morphospace occupation and extinction events. Therefore, it is prudent to explore patterns of morphological diversity using a combination of metrics that capture different aspects of evolutionary change. However, many of the metrics tested depend on distance measures in a morphospace that has affine rather than metric properties (Mitteroecker and Huttegger, 2009). Therefore, utilizing a combination of metrics that capture different aspects of evolutionary change and avoid mathematical issues such as the ratio of generalized variance, (Huttegger and Mitteroecker, 2011), preordination distance (Deline et al., 2020), and the rate of character change (Hopkins and Smith, 2015) can produce more meaningful results.

## Comparisons between Methods

The degree to which the different methodologies commonly used to quantify morphology produce congruent results is vital in interpreting the overarching patterns of disparity. If these studies are all estimating similar trends in the morphology that is inherently linked to the underlying phylogenetic structure, then broader conclusions can be reached through meta-analyses of disparity studies (such as Hughes et al., 2013). Villier and Eble (2004) explored this by comparing the patterns of disparity of spatangoid echinoids resulting from a landmark geometric morphometric and discrete character-based methodologies. The resulting trends were fairly consistent and mostly differed in the magnitude of change between time bins rather than the overall pattern. An optimistic interpretation of this result is that the different methodologies are all capturing a similar underlying morphological signal, whereas the differences are based on capturing slightly different aspects of overall form. Studies on other taxonomic groups have largely found the same relationship between different methodologies (e.g. Hetherington et al., 2015; Schaeffer et al., 2019). However, the relationship may be more complicated (e.g. Mongiardino Koch et al., 2017). A recent study in Paleozoic fish showed similar overall distributions of taxa within morphospace and disparity in time using character-based and geometric methodologies (Ferrón et al., 2020). However, this study also showed disagreement in terms of the relative disparity between subgroups, highlighting the different aspects of captured morphology. Whether the disagreement between metrics arises from differing sensitivities to convergent evolution or environmental distributions is unclear in this case, but more study is needed to better understand the observed patterns, especially those in which only character-based approaches are available, such as those exploring disparity at larger taxonomic scales.

## 3 A priori Choices in Experimental Design

A priori choices in experimental design have significant effects on the perceived patterns of morphological evolution through time. This can include the overall methodology, the level of detail being collected, the taxonomic breadth of study, as well as the taxonomic, geographic, and temporal scale utilized. In addition, different community structures can lead to increased exclusion of rare taxa from analyses potentially skewing the resulting patterns. Finally, taphonomic processes can potentially alter perceptions of morphological diversity through time by the loss or alteration of features. Subsequently, there will be a focus on character-based studies in that they provide the opportunity to explore many different and potentially broad taxonomic groups. In addition, character-based studies can appear at face value to be more applicable and straightforward (e.g. utilizing pre-existing cladistic character data), but they often require significant thought in experimental design.

### Volume and Type of Characters

One of the first choices in experimental design is deciding the scope and detail needed to address a particular hypothesis. If morphology is quantified too broadly, or the taxonomic coverage is too narrow, the study might be inadequate to answer the scientific question being asked. However, increasing the level of detail captured in the morphological analysis, or raising the taxonomic coverage, can potentially and exponentially increase the time and effort required to collect the relevant data. Therefore, it is paramount to explore how much information is required to quantify morphology in a meaningful manner in terms of the number of specimens, the number of taxa, or the size of the character matrix prior to beginning a study.

Deline and Ausich (2017) constructed a novel character suite to describe Paleozoic crinoids that approximately doubled the number of characters

(178 vs 92) and character states (666 vs 322) used by Foote (1999). Even though the overall sizes of the data sets were dramatically different, the resulting morphospaces were statistically similar in structure (Mantel test, p=0.001). The increase in detail allowed the clarification and differentiation of certain aspects of morphology (e.g. clearly differentiating cladids and flexibles and capturing the uniqueness of the calceocrinids), but the level of similarity was striking. To explore this further, Deline and Ausich (2017) ran a series of character-based rarefaction analyses, in which they randomly subsampled between 5 and 178 out of 178 total characters without replacement 1,000 times for the newly constructed crinoid data set. An ordination was conducted for each subsample and multiple metrics were calculated to determine the number of characters needed to accurately characterize a particular aspect of crinoid morphology. A relatively small number of characters (~20 per cent, 35 characters) could be used to capture the relevant morphologic properties of the entire data set, such as the number of statistically distinctive clusters, overall variance, fidelity of the morphospace to the original data set, and relative disparity of subgroups (Figure 3A–C). The low number of characters required to quantify morphology is likely the result of correlated characters, such that even though they capture distinctive and unique aspects of anatomy, they are redundant (i.e. vary in similar ways between crinoid groups). For example, the presence and properties of fixed rays and tegmens both largely distinguish camerate from non-camerate crinoids. The inclusion of these characters builds increased separation between the groups within morphospace, but the increased degree of separation is not required to capture the overall properties of the morphospace. These results are similar for different subsets of the data, but more characters are required to capture the morphological signals when the subgroups have a large degree of morphological overlap (e.g. monbathrid and diplobathrid camerates). This study highlights that a small number of characters can potentially capture the morphological properties of a disparate group. Obviously, the number of characters required would vary with different taxonomic groups, but this study provides a methodology for conducting a pilot analysis to determine the amount of data needed to capture a particular trend.

In many organisms, a particular region is better preserved or considered more important taxonomically or ecologically and, thus, it is over emphasized in studies of overall morphology. Deline and Ausich (2017) used similar methods to those described previously to explore how relative disparity of crinoid subgroups differed with shifting emphasis on different aspects of morphology. They compared random character reduction to subsequently reducing the

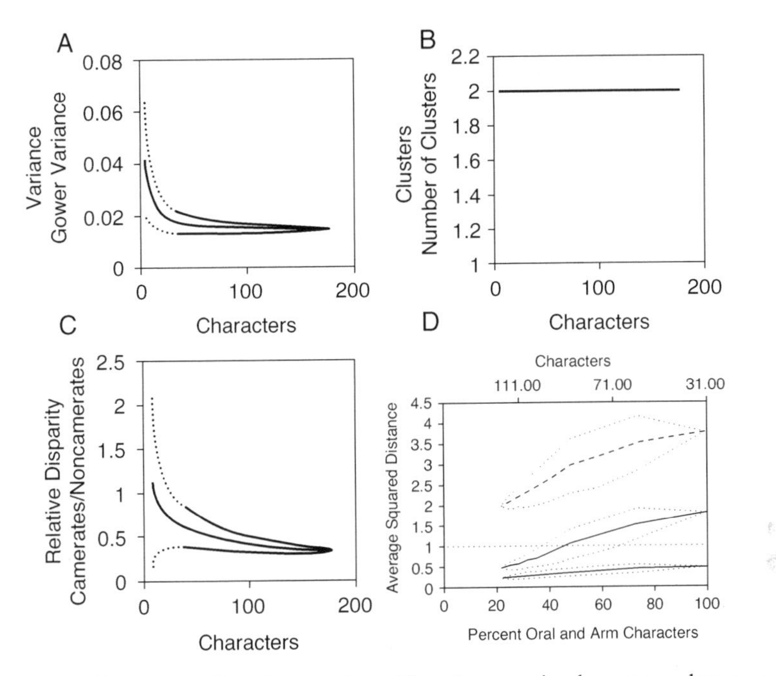


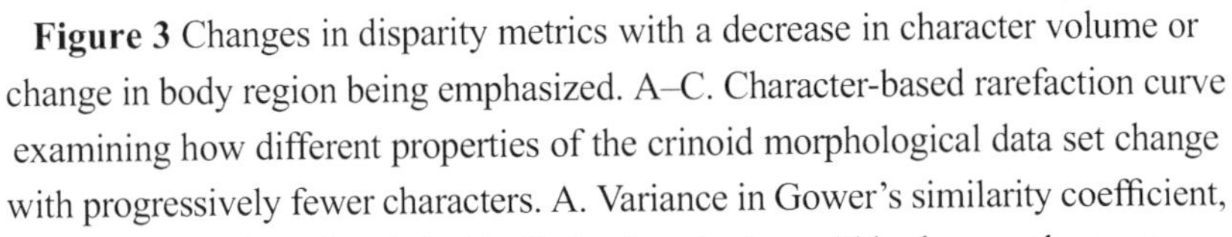

**Figure 3** Changes in disparity metrics with a decrease in character volume or change in body region being emphasized. A–C. Character-based rarefaction curve examining how different properties of the crinoid morphological data set change with progressively fewer characters. A. Variance in Gower's similarity coefficient, B. The number of statistically distinctive clusters within the morphospace. C. Relative disparity between camerate and cladid crinoids. D. Relative disparity of different crinoid groups with a progressive de-emphasis on oral surface and arm characters. Dashed line: disparids/cladids; black line: camerates/cladids; gray line: camerates/disparids. Error envelopes based on standard error of 1,000 replicate samples. Figures modified from Deline and Ausich 2017.

number of characters in a particular body region while retaining those of other regions. The results varied dramatically depending on the body region being examined and the taxa being considered. For instance, the ratio of disparity between camerates and cladids remained consistent with the de-emphasis of characters associated with the periproct (e.g. anal ray plates, anal sac, and anal tubes). In other cases, reducing the number of characters from a certain body region shifts the relative disparity between two groups or changes which group appears more disparate. For instance, reducing the number of the oral surface and arms characters by 50 per cent switched whether cladids or camerates appeared more disparate (Figure 3D). This means that a shifting perspective on which aspects of morphology are relevant (denoted by more characters) can significantly bias our view of morphology. This is especially important with the utilization of cladistic characters for studying disparity in which preference for particular homology schemes focusing on a particular body region can greatly influence the resulting patterns.

## Binning and Scale of Study

In addition to the types of characters that are used, a priori choices need to be made regarding the temporal binning and scale of the data being collected. Selecting the taxonomic-level of the coded organisms can potentially alter the perceived trends. As the chosen operational taxonomic unit becomes broader, the meaning of disparity can also become murkier. The amount of morphologic variation contained within a species clearly differs depending on the taxa and author, but these differences in systematic practice (i.e. splitting verses lumping taxa) can increase with taxonomic level such that the empirical meaning of the comparisons becomes dramatically inconsistent. However, attempting to limit this bias by exclusively using species-level rank can severely limit the scope of the study. Alternatively, species ranges are shorter and might therefore be more susceptible to sampling biases than using genus ranges (Villier and Eble, 2004). Several studies have explored how the choice in operational taxonomic units can alter trends in disparity through time. Villier and Eble (2004) examined disparity trends at both the species- and genus-level in spatangoid echinoids and found clear congruence. However, they note a strong correlation between the number of species and genera. This correlation, along with the limited temporal range and taxonomic scale, likely plays a role in the similarity in trends. In effect, a taxonomically narrower study is likely to be composed of organisms that were described under similar systematic practices. However, times of rapid speciation can dramatically increase the number of species per genus (e.g. the Great Ordovician Biodiversity Event), which could then alter the amount of disparity contained within a genus. Therefore, in broader studies, the number of species per genus could vary non-randomly and produce a disconnection between genus- and species-level patterns. Deline et al. (2012) explored the relationship between genus and species disparity patterns in early Paleozoic crinoids and found a non-systematic disagreement in the trends in disparity. As crinoid diversity expanded through the Ordovician into the Silurian, there was a decrease in single-species genera and an overall increase in the number of species per genus. Therefore, there was strong agreement between disparity patterns initially but diverging through time. The genus-level pattern captured an increase in disparity in the Early Silurian, which wasn't apparent from an examination of species. It would then be expected that the meaning of disparity at higher taxonomic levels would become more complicated. Broader patterns could simply be an amalgamation of the disparity and evolutionary patterns at lower taxonomic levels or the morphological patterns at the origin of larger groups could indicate unique evolutionary mechanisms. Hopkins and Smith (2015) explored disparity at the 'class' and 'subclass' levels in post-Paleozoic echinoids. At the class level, echinoid disparity was fairly consistent through time, with several relatively short peaks of morphological change. These peaks represented early bursts of evolutionary innovation within clades of irregular echinoids. Thus, the broad patterns in disparity were controlled by a baseline of character change within established clades and the appearance of new characters at the origin of new clades. Whether this pattern is consistent with other groups or during other intervals (e.g. the early evolution within a phylum) needs to be further examined.

Morphological patterns can be driven by the selective extinction and expansion of specific regions of morphospace, which is likely the result of ecological interactions such as competition and predation or environmental specialization. However, these interactions are local interactions, whereas disparity is most often explored as an expansive global pattern within a clade. Therefore, just as it is valuable to explore biodiversity at the local (alpha) and global levels (gamma), the geographic scale of analysis can also be valuable to explore in studies of disparity. Deline et al. (2012) explored the relationship between local (i.e. within a biofacies) and regional disparity in early Paleozoic crinoids. Laurentian crinoid biodiversity during the early Paleozoic showed similar patterns both within biofacies and regionally with a large increase in diversity during the Middle to Late Ordovician, a drop at the end of the Ordovician, then a recovery during the Early Silurian (Deline et al., 2012). Conversely, there was little change in within-biofacies disparity even with changes in regional disparity throughout this interval. The lack of change within individual biofacies was largely the result of two factors. First, most crinoid occurrences contained representatives from the four major subclasses (Figure 1), which established a baseline of disparity contained within each biofacies. Second, the morphology within each biofacies was often regulated by the environment (e.g. energy level), such that the disparity in each of the subclasses was fairly low. Therefore, if the distinctiveness of the subclasses remained consistent with little within-cluster variance, the overall biofacies disparity was static. Again, this shows the choice of geographic scale from local to regional can allow the exploration of different aspects of morphological evolution.

In addition to the choices regarding character type, character volume, taxonomic scale, and geographic scale, the coarseness of temporal binning can alter patterns of disparity through time. A courser binning scheme may potentially reduce uncertainty in species ranges but would also average out and dampen the resolution of the patterns of disparity through time. Alternatively, too fine a binning scheme could result in bins that have small sample sizes or are empty. In most cases, temporal binning is based on commonly used stratigraphic scales such as stages or biostratigraphic zones. This choice allows an easy comparison to other data that often use a similar temporal binning scheme. However, this choice also creates bins that are unequal in duration, giving a skewed view of change and potentially make false assumptions regarding models of evolution (Guillerme and Cooper, 2018). Villier and Eble (2004) compared disparity patterns in Cretaceous spatangoids using multiple temporal bins including a stage-level analysis as well as breaking the study interval into five bins that were roughly equal in duration. As was expected, the general pattern was consistent in both binning schemes, but the stage-level analyses provided more resolution and patterns that were not apparent in the coarser analysis (Villier and Eble, 2004).

Guillerme and Cooper (2018) explored this issue in more detail using data from multiple disparity studies including Wright's (2017b) study focusing on eucladid crinoids. In this study (Guillerme and Cooper, 2018), they proposed a method called time-slicing in which disparity is sampled from a time-standardized phylogenetic tree at multiple fixed points. The use of a phylogenetic tree allows the modelling and sampling of ancestral character states based on multiple models of evolutionary change (e.g. punctuated or gradual). This produces multiple snap shots of disparity rather than an average amount of change within a particular time interval. This approach was then compared to binning based on stratigraphic intervals as well as bins of equal duration. They did not find any systematic differences in estimations of disparity through time between using time bins and time slices (Guillerme and Cooper, 2018). However, there can be differences in relative disparity, the timing and duration of disparity peaks, and the effects of mass extinctions. Therefore, there can be an effect on the biologically important aspects of disparity analyses such that it is prudent to explore the data utilizing multiple methods preferably beyond exclusively time binning (Guillerme and Cooper, 2018).

## 4 Taphonomy and Missing Data

The loss of information from taphonomic degradation can potentially have severe effects on studies of morphological evolution. These effects can include an increase in the amount of missing data, the preferential exclusion of rare taxa, and the non-random loss of morphological features. Echinoderms are mainly constructed of variably articulated ossicles that disassociate following death. The rate at which disarticulation occurs varies dramatically across the phylum as well as within different environments (Brett et al., 1997). Therefore, echinoderms display a wide range of preservational modes such that they are ideal for testing how preservation can alter trends in disparity.

The foremost issue caused by taphonomic processes in the study of disparity is an increase in missing data within the analysis. In character-based studies, missing data is often handled by excluding that particular missing character from the calculation of distance between two taxa. Therefore, if taxa all contain some degree of missing information then the distances between taxa are calculated by different subsets of characters resulting in inconsistencies in the resulting pairwise distance matrix (Gerber, 2019). These inconstancies can in practice result in triangular inequalities and negative eigenvectors. There are multiple methods to correct for this mathematical issue (e.g. Cailliez, 1983), but the missing data has a broader effect on the resulting ordination regardless of the corrections. Gerber (2019) illustrates this issue using simulated data in which he compares the correlation between the original distance matrix and the Euclidean distance between taxa within the resulting morphospace degraded with increasing levels of missing data. He found that as the proportion of missing data increased, there was a linear decrease in the rank order correlation (slope = –1). However, even though rank order of the distances changed significantly, the mean pairwise distance remained consistent (Gerber, 2019). This may sound encouraging; however, it is the result of the random loss of missing data, which is unlikely with taphonomic degradation. Taphonomic processes produce a systematic and non-random loss of data, which is likely to increase or decrease disparity compared to taphonomically unaltered data. Therefore, an increase in missing data from lower rates of preservation can fundamentally alter perceived patterns of disparity through time. The relationship between missing data and disparity studies was also examined by Lloyd (2016) by comparing the consistency of distance matrices with an increase in missing data depending on the chosen coefficient and character type. Lloyd's (2016) results are mirrored in those of Gerber (2019) in which most of the distance metrics follow a linear trend with an increase in missing data leading to a decrease in fidelity. At very high levels of missing data, all of the metrics have low fidelity, as is expected. However, the relative performance of different distance metrics shifts from no missing data to high levels of missing data. Therefore, the selection of distance metric should depend on the amount of missing data and, thus, hopefully lessen the confounding effects of that missing information (Lloyd, 2016).

With the understanding that missing data can substantially reduce the reliability of the ensuing distance matrix and morphospace, multiple methods have been suggested to avoid this potential problem. Gerber (2019) suggests culling characters and taxa with high levels of missing data or studying taxa based on smaller subsets to reduce the overall percentage of missing data. Smith et al. (2014) highlighted that one potential issue of missing data is in the non-random distribution through taxa being examined, especially when comparing fossil and modern organisms. They suggested that since many characters are correlated, that missing data could be potentially added to the analysis in a way that would reduce fidelity less than random character loss (Smith et al., 2014). That could potentially normalize the distribution of missing data through the analysis, without dramatically altering the results. However, this method is still adding uncertainty into an analysis and might be inappropriate for an analysis with an already high proportion of missing information. Alternatively, Brusatte et al. (2011) proposed a method to reduce the proportion of missing data by modelling missing character states based on phylogenetic position. This can be done in a similar method as reconstructing ancestral states using parsimony, maximum likelihood, or Bayesian methodologies. The results of the reconstructed character states are contingent on the tree topology as well as the model of evolutionary change. Therefore, a resolved phylogenetic hypothesis is required for reliable results. In addition, exploring the effect of the use of different evolutionary models or trees in character reconstruction is prudent. Finally, using a Bayesian approach, such as stochastic character mapping (Huelsenbeck et al., 2003) allows running thousands of simulations to establish the posterior probability of missing character states. This potentially allows the reconstruction of missing data in which a character state is probable, while leaving less certain states as missing. This procedure could reduce the issues of missing data, without adding undue uncertainty into the analysis.

Deline and Thomka (2017) explored the potential distribution of missing data within echinoderm data sets and its effect on studies of disparity. Using crinoid and blastozoan data sets, they identified characters likely to be lost during progressive taphonomic alteration based on the work of Brett et al. (1997). Using four distinctive taphonomic grades, they compared the fidelity of the both random and taphonomic character loss to the original distance matrix. For blastozoans, the taphonomic loss of information mimicked random character loss (Figure 4A), which is likely the result of the loss of redundant characters and the number of characters that can be identified with isolated elements (e.g. respiratory structures). In addition, many blastozoan groups lack set patterns in thecal plating such that disarticulation does not dramatically alter the characters chosen to quantify morphology. The pattern for crinoids is very different

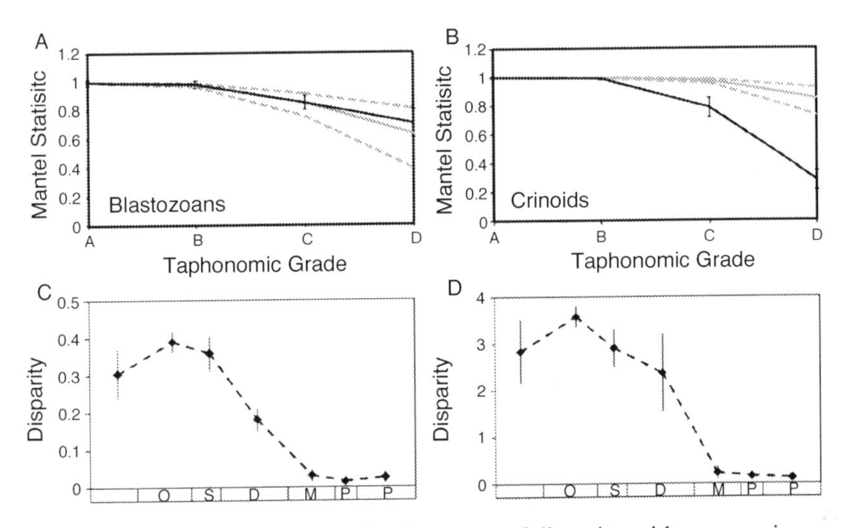


**Figure 4** Changes in the perceived patterns of disparity with progressive taphonomic degradation. A, B. Fidelity measured with the Mantel Statistic comparing the Gower dissimilarity coefficient of taphonomically altered and original data set (black line). The loss of fidelity can be compared with random character loss shown in grey. C, D. Shifts in blastozoan disparity through time without (C) and with (D) exaggerated taphonomic degradation. Figures modified from Deline and Thomka 2017.

(Figure 4B): the taphonomic loss of characters dramatically reduces fidelity compared to random character loss. Crinoids appear much more taphonomically sensitive because many characters require intact portions or entire calyxes to accurately code. Quantifying crinoid morphology is reliant on the relative position, number, and structure of calyx plating rather than the characteristics of individual plates such that the loss of those characters greatly alters the resulting distance matrix effectively blurring the major morphological groupings within crinoids (Deline and Thomka, 2017). Therefore, the taphonomic effects on a data set are also reliant on the type of characters chosen to quantify morphology such that consideration for preservational differences should be taken into account in selecting characters. In other words, before constructing a character suite to quantify morphological trends, both the morphology as well as the taphonomy should be closely examined. To further explore the role of taphonomy on estimating patterns of disparity, Deline and Thomka (2017) exaggerated the taphonomic alteration in the blastozoan data set (Foote, 1992) unequally through time based on the frequency of Lagerstätte. Increasing the taphonomic degradation within the data set had little effect on the overall trend in disparity through time (Figure 4C, Dd), differing largely based on the shift of a single blastoid species into an outlier position in morphospace. Whether this result is further illustrating Gerber's (2019) point that increasing the frequency of missing data can greatly degrade the data without altering patterns of disparity through time or disparity patterns can be recovered despite taphonomic degradation needs to be further explored.

Finally, taphonomic processes can remove features from known organisms but can also prevent taxa from being recovered at all. This taphonomic removal is not random and depends on an animal's morphology, the environment in which they lived, and the abundance of the organism. There is a strong taphonomic bias against the preservation of rare taxa in the fossil record. Therefore, unrecovered rare taxa could additionally alter patterns of disparity, especially since rates of preservation are not uniform through time. Deline (2009) examined the relative disparity (Foote, 1993) of crinoids depending on their abundance within well preserved Ordovician assemblages. Deline (2009) found no statistical difference in the average partial disparity between rare and common crinoids within those assemblages. Even though rare taxa are often considered morphologically aberrant, this does not seem to hold true and the exclusion of rare taxa might represent a random rather than confounding bias through time.

Overall, there are many choices and potential biases that can alter perceived trends in disparity through time. Therefore, the methodologies, metrics, and potential biases should be considered in the experimental design in order to better interpret the resulting patterns of morphological evolution through time.

## 5 Patterns in Phanerozoic Disparity

Despite potential biases and issues with missing data, several patterns have been repeatedly recognized regardless of the method utilized, which suggests these trends might reflect shared underlying patterns in macroevolution. In particular, many studies have documented common patterns in disparity during the origination of a clade, during biotic turnover events, and during the later stages in a clade's history in which constraints in morphology arise.

Clades are recognized in hindsight, such that they are defined largely by their success and subsequent diversification following their origination. With this perspective on the origination of clades, a common morphological pattern is expected. Clades originate with an evolutionary innovation, extinction of competitors, shift into a new ecological niche, or migration into a new geographic area. Following this initiating event, there is the potential for diversification within that new space, leading to an increase in disparity. Eventually, disparity will slow as the new ecological space fills and competition increases. Finally, the morphological diversity will stagnate until something changes (e.g.

ecological, biogeographic, or environmental) or the clade becomes extinct. Obviously, these patterns could change through time depending on the size of the clade and the underlying developmental patterns. Hughes et al. (2013) compiled a metanalysis of 98 metazoan clades to test for the presence of this overall pattern in the disparity history of a clade. They explored the overall history of disparity within a clade by calculating the morphological centre of gravity which measures the timing of morphological expansion and peaks in disparity in comparison to the clade's overall duration. If a clade follows the hypothesized pattern discussed previously, it would have a centre of gravity lower than 0.5, such that the disparity would be considered bottom heavy. However, if disparity within a group steady expanded such that peak disparity occurred later in a clade's history the centre of gravity would be higher than 0.5 and be considered top heavy. There was no overall trend through the Phanerozoic in the centre of gravity within a clade, but several important trends could be observed. First, if a clade was 'prematurely' truncated at a mass extinction, it was likely to have a top-heavy pattern of disparity. Whether their disparity profile would have differed given a continued opportunity is unknown such that those clades have little bearing on the above hypothesized pattern. However, clades that do not terminate at a mass extinction (63) are three times more likely to have a bottom-heavy than a top-heavy centre of gravity (Hughes et al., 2013). Hughes and colleagues also found a paucity of clades originating immediately following mass extinctions that had a bottom-heavy pattern of disparity, which can be interpreted in multiple ways. First, this could reflect a pull of the modern if a clade originates following a mass extinction, which could result in less time to follow the predicted pattern and would more likely be truncated. In addition, the opportunity following a mass extinction, especially for more minor or more selective mass extinctions, may be muted compared to that at the beginning of the Paleozoic. Alternatively, the decrease in opportunity could be the result of increasing developmental constraints through time, even with ecological opportunity.

Several studies on echinoderms, particularly crinoids, help to elucidate this pattern given their apparent early Paleozoic diversity in form. Foote (1994a) examined the early morphological history of crinoids. Crinoids have erect ambulacra that are used as efficient feeding structures and following this morphologic innovation their biodiversity increased into the middle Paleozoic. However, despite the continued taxonomic diversification, their disparity peaks in the early Late Ordovician as measured by the variance and the number of realized character combinations. Foote (1994a) interpreted this pattern as an early exploration of major crinoid forms during the Ordovician followed by subsequent exploration around those forms during later diversification. Wright (2017b) further explored eucladid crinoid disparity patterns within a phylogenetic context to examine rates of character change through time. Using this alternative approach, Wright (2017b) found that the early eucladids had elevated rates of character change (Figure 5A) consistent with rapid morphological exploration following the evolutionary innovation at the origin of the clade. The rates reported by Wright (2017b) following this initial burst dramatically decreased consistently with increased constraint either from saturating the potential niche space or through the loss of developmental flexibility. The early peak in disparity within crinoids found by Foote (1994a) and Wright (2017b) is consistent with the results of Deline et al. (2020) for echinoderms as a phylum. Deline et al. (2020) found an increase in variance through the Cambrian and peaking in the Late Cambrian to Early Ordovician. Therefore, for echinoderm clades originating during the early Paleozoic there is a consistent pattern of an early and rapid expansion of morphological features in spite of the scale of study or the method being used.

Following the initial diversification, multiple studies found long periods of static disparity in which the rate of character change was low (Foote, 1999; Wright, 2017b). This damping of morphological innovation later in a clade's history has largely been attributed to the competition within a particular niche from ecological incumbents coupled with stabilizing selection and an increase in developmental constraints. However, these constraints do not appear to be consistent through the rest of the clade's history. Foote (1996) examined trends in crinoid disparity across the Permian mass extinction. Crinoids were mostly decimated during this interval with only a few taxa surviving. In the early Triassic, crinoid disparity underwent a rapid diversification showing their continued morphological plasticity with the ecological release even with some level of incumbent competition (Foote, 1996). Wright (2017b) found an additional burst in morphological change within eucladids much later in the clade's history within the Late Pennsylvanian (Figure 5A). Interestingly, this peak does not coincide with an increase in disparity highlighting the need for exploring morphologic evolution with multiple methods. Furthermore, this peak is unrelated to the appearance of novel features or the appearance of a new subclade, rather it is the result of elevated rates of change within multiple subgroups during an interval of environmental change (Wright, 2017b). Again, this highlights the continued ability for morphological change in echinoderms following the initial burst of disparity at the origin of the clade.

Romano et al. (2018) explored the disparity of a group of crinoids that originated much later in the Phanerozoic, cyrtocrinids. Crytocrinids were a dominant crinoid group during the Jurassic and Cretaceous with paedamorphic features including a simplified cup and arms. This simplification likely

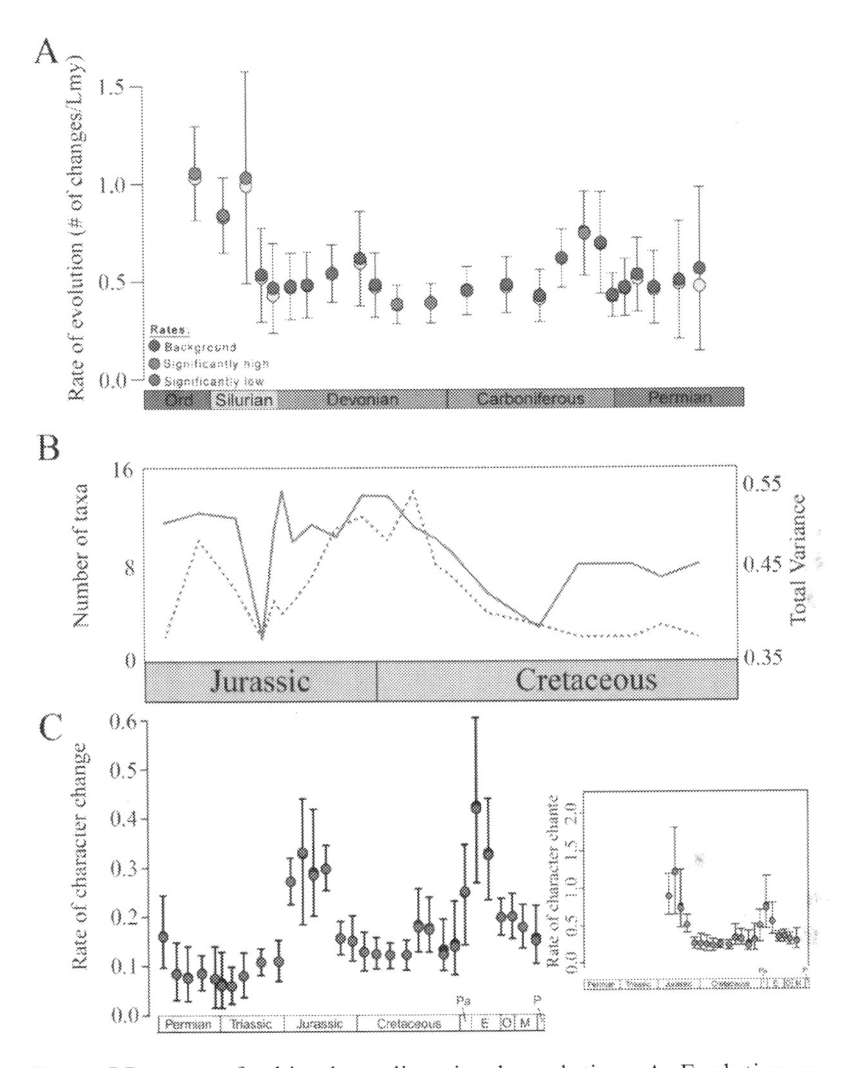


**Figure 5** Patterns of echinoderm disparity through time. A. Evolutionary rate of morphological character change in Paleozoic eucladid crinoids modified from Wright 2017b. Patterns derived from the maximum-likelihood analysis of time-calibrated trees from the Bayesian posterior distribution. B. Disparity (red) and diversity (blue) in cyrtocrinid crinoids modified from Romano et al. 2018. Disparity calculated as total variance. C. Rate of morphological evolution in echinoids from Hopkins and Smith 2015. Rates calculated as the mean number of character changes per 10-million-year time interval accounting for phylogenetic uncertainty.

gave the group a selective advantage during environmental changes within the Tethys basin (Romano et al., 2018). In this case, the evolutionary innovation within cyrtocrinids was the result of a shift in development rather than the appearance of a novel feature. This mirrors the results of Wright (2017b) in regard to Pennsylvanian eucladids in which simplification and developmental shifts resulted in elevated rates of morphological change. Even without the appearance of multiple novel characteristics, cyrtocrinids follow the same bottom-heavy pattern of disparity through its duration (Figure 5B). This highlights the continued opportunity within crinoid clades for morphological change even following the initial establishment of body plans. This also suggests significant complexity within developmental constraint in which the opportunity for new features is potentially reduced, but developmental processes remain flexible allowing later bursts of innovation.

The pattern shown in crinoids is likely not an anomaly within echinoderms as similar patterns have been observed within echinoids. Eble (2000) examined patterns of morphological diversity within Mesozoic atelostome echinoids. The clade as a whole increased in disparity through the Jurassic then stagnated during the Cretaceous following the trajectory of a bottom-heavy disparity profile (Eble, 2000). The rise in disparity was the result of the progressive appearance of different subclades such as the holasteroids and spatangoids that collectively boosted the disparity of the entire group. However, after the appearance of the individual clades, disparity becomes static as the within-group disparity stagnates or decreases. Hopkins and Smith (2015) broadened the work of Eble (2000) by examining morphologic trends and evolutionary rates in all post-Paleozoic echinoids. Hopkins and Smith (2015) found a steady and low rate of character change within echinoids during the last 300 million years, punctuated by elevated rates in the Jurassic and Eocene (Figure 5C). Echinoids originated in the Ordovician (Smith and Savill, 2001), such that the study interval was far beyond any initial burst of morphological innovation. Therefore, 250 million years after the clade's origination, low rates of morphological change are expected based on the proposed patterns described previously. The elevated peaks coincide with significant ecological shifts towards grazing and gravity-sieving habits in irregular echinoids (Hopkins and Smith, 2015). The morphological innovations associated with these new ecological modes spiked rates of change, but the elevated rate of change was relatively short lived such that it soon normalized to the previous background levels. Hopkins and Smith also point out that the elevated evolutionary rates within irregular echinoids is closely related to a shift in growth from plate addition to plate accretion. This ontogenetic shift and associated change in developmental processes again highlights the continued morphologic flexibility within echinoderms. In addition, the appearance of new irregular echinoids coincides with both changes in developmental patterns as well as the appearance of novel features. Obviously, the loss of clade-level disparity within echinoderms through the Phanerozoic corresponds to a contraction of morphologic diversity within echinoderms. However, the appearance of post-Paleozoic clades (e.g. cyrtocrinids or clypeasteroids) suggests a prolonged potential for morphological innovation within the phylum and developmental flexibility far beyond the initial morphological diversification in the Cambrian and Ordovician.

The general shape and distribution of taxa within morphospace as well as the patterns of disparity through time have led paleontologists to attempt to understand the developmental patterns underlying large-scale morphological change. These patterns, paired with modern exploration of development within extant organisms, can potentially provide insight into the mechanisms leading to the prevalent patterns discussed. For instance, the initial clumpy distribution of taxa in morphological space has been noted in many instances and could be the result of multiple mechanisms (Erwin, 2007). These could include the extinction of intermediary forms, contingency and the chance extinction of early forms, an incomplete exploration of the myriad available forms, occupation of multiple fitness peaks, or deep and highly structured developmental processes within a larger clade (Erwin, 2007). Differentiating between these hypotheses for the observed pattern is difficult with incomplete preservation and the lack of knowledge of developmental processes in deep time. Given aspects of genotype and development are known only from extant taxa, there is a 500-million-year gap in knowledge in reference to the processes occurring at the origin of a clade. However, the developmental patterns resulting in the spatial appearance and distribution of features can be inferred based on extant taxa (Salazar-Ciudad, 2006), such that patterns of morphology can be captured with a developmental model in mind. Overall, multiple attempts have been made to better explore the developmental aspects of disparity through time.

Ciampaglio (2002) attempted to explore the connection between disparity and developmental patterns based on the previously discussed morphological data sets of crinoids (Foote, 1999) and blastozoans (Foote, 1992). To test the role of ecology compared with developmental constraints, Ciampaglio (2002) compared the relative disparity before and after extinction events. Based on the assumption of permissive ecology following an extinction, a return of disparity to pre-extinction levels would imply that developmental constraints are not becoming more rigid through time. In most cases, he found disparity rebounded rapidly following extinction to comparable levels in both crinoids and blastozoans. Furthermore, Ciampaglio (2002) classified characters as primarily ecological, developmental, or some combination thereof. He defined ecological characters as those that capture an aspect of morphology that interacts with some external aspect of ecology such as hydrodynamic regime, feeding, or metabolism, whereas developmental characters were defined as features that are not directly linked to ecology but instead are more constructional in nature. The issue in this approach is that all characters are formed through some combination of developmental processes such that splitting characters on this basis is somewhat arbitrary. However, this study does highlight the continued morphological plasticity found within echinoderms.

Gerber et al. (2011) explored the relationship between development and disparity from a theoretical prospective. They used a simple model to explore the relationship between juvenile disparity and adult disparity. In addition, they explored how developmental changes in growth could alter both the adult morphology and, therefore, trends in disparity through time. Most importantly, comparing adult disparity, juvenile disparity, and the trajectory of change between them allowed Gerber et al. (2011) to distinguish different types of developmental shifts (e.g. ontogenetic scaling or lateral shifts in morphology). However, in most fossil echinoderms there is limited information regarding ontogeny. In addition, most ontogeny studies on echinoderms focus on more derived taxa that occur later in a clade's history or where the earliest ontogenetic stages are missing (e.g. Brower, 1974; Waters et al., 1985), but this approach could be appropriate in certain systems (e.g. Zamora et al., 2013). Nevertheless, this study provides an analytical framework for exploring developmental shifts in fossil organisms.

Deline et al. (2020) further examined the developmental mechanisms that occurred at the origin of echinoderms. In this study, they constructed an expansive character matrix capturing the morphology found within early Paleozoic echinoderms. Deline et al. (2020) structured this character matrix hierarchically such that characters built upon each other, mimicking the complexity of biological features. Riedl (1977) proposed a model of morphologic evolution in which large-scale characters become more complicated through time, essentially 'burdened' with their own complexity. He proposed that these characters are more likely to become fixed and, thus, resistant to change through time. This could also lead to the patchy distribution of clades through time as has been observed in many studies of morphological disparity. This model has been further expanded based on recent discoveries in developmental gene regulatory networks (GRNs). These networks have a complicated hierarchical structure with certain elements relatively impervious to evolutionary change (Davidson and Erwin, 2006). These properties of GRNs have been recognized in modern echinoderms for features such as the biomineralization of calcitic stereom, which has been conserved across a broad taxonomic distance (Erkenbrack and Thompson, 2019). Furthermore, the appearance and fixing of these GRNs has been hypothesized to underlie the early origin of animal body plans (Davidson and Erwin, 2006). Deline et al. (2020) compared the phylogenetic signal of characters with their burden (e.g. the number of contingent characters) and found no significant relationship. Essentially, the ability for a character to change is independent of its burden or complexity. This allows a substantial and prolonged ability within echinoderms for evolutionary innovation, as was observed in post-Paleozoic echinoids and crinoids. There are multiple examples within Paleozoic echinoderms of this plasticity, such as the repeated loss of stem within pelmatozoan groups (e.g. diploporitans or cytrocrinids), loss of erect feeding appendages in lieu of recumbent ambulacra (e.g. hybocystids), and the decalcification of the oral frame (e.g. camerate crinoids). Overall, this developmental perspective on disparity shows the continued ability to repurpose and break constraints which prevents the restriction of morphological diversifications through time (Budd, 2006).

## 6 Future Directions

As has been discussed, there has been a rich history of study regarding the morphological evolution and trends in disparity in echinoderms through time. However, there are ample opportunities for further study and exploration. Foremost, the combination of phylogenetic hypotheses, along with morphology to construct phylomorphospaces, allows for potential insights into a number of different evolutionary processes. This is particularly promising given the plethora of recent examinations of fossil echinoderms using modern phylogenetic practices (e.g. Bauer et al., 2019; Cole, 2018; Sheffield and Sumrall, 2019; Wright, 2017a). Capturing both the range of morphologies along with the pathway of how those morphologies evolved provides insights into the rate and direction of change rather than just the general spread and location of taxa within morphospace (Lloyd, 2018). Exploring the rate of change allows the testing of models to better understand the underlying evolutionary mechanisms. Establishing the directional vectors of change in morphospace allows an easy test of convergent evolution or shifts in evolution at particular intervals in the phylogenetic tree.

There are multiple methods to construct phylomorphospaces, including both pre- and post-ordination methods. Given a tree with known branch lengths and the distribution of taxa in a morphospace, a position of an ancestral node can be calculated using maximum likelihood (Sidlauskas, 2008). This methodology is straightforward and easy to compute but has some significant drawbacks as discussed by Lloyd (2018). First, this methodology forces ancestors into the area already explored by the tip data, thus assuming the sampled taxa are morphologically extreme whereas the ancestors are more average in form

(Lloyd, 2018). Second, since ordinations of character data are nonmetric as discussed previously, then not all positions across morphospace are possible such that the placement of an ad hoc ancestral node is problematic. Essentially, there is a disconnect between the placement of the ancestor in morphospace and the likely characters that would represent that position. Finally, the post-ordination methods of phylomorphospace construction impose a strong phylogenetic signal onto the resulting structure (Lloyd, 2018). Obviously, morphology is a by-product of the phylogenetic relationships but in most instances an overt signal is to be avoided in studies of disparity. However, these methods can be utilized to visualize broad evolutionary patterns, with the caveat that the reading of the patterns is done with an understanding of their limitations. These methods were used in the previously discussed Hopkins and Smith (2015) and Wright (2017b) studies (Figure 6). However, in both cases, the authors note that the phylomorphospaces are for visualization of the phylogenetic structure in morphospace rather than for making an inference regarding the underlying morphologic characteristics at a particular ancestral node.

Alternatively, pre-ordination methods utilize a tree with known branch lengths, the characteristics of the examined taxa, and a model of evolution to estimate ancestral character states. This methodology can minimize the phylogenetic signal in the phylomorphospace and provide more realistic estimations of the phylogenetic signal, rates of character change, and degree of convergent evolution (Lloyd, 2018). Therefore, if the phylomorphospace analyses are being utilized to identify evolutionary processes, pre-ordination methods are preferable though still not free of biases, which are inherent to any form of inference. Deline et al. (2020) estimated ancestral character states in the construction of an early Paleozoic echinoderm phylomorphospace (Figure 6). This was done using stochastic character mapping (Huelsenbeck et al., 2003), which is a Bayesian approach that stochastically models character evolution through a phylogenetic tree. Running repeated simulations allows the calculation of probabilities at each node, such that the most likely ancestral character state and degree of certainty can be accurately assessed. Lloyd (2018) still cautions that these types of methods are still influenced by the underlying phylogenetic signal and are further distorted through ordination. Nevertheless, these methods provide a powerful tool in exploring evolutionary change in a more in-depth manner. This methodology also provides the opportunity to explore or re-explore patterns of morphological evolution within many echinoderm groups. It enables a comparison of the rate of change alongside the breadth of morphology (e.g. Wright, 2017b) and can provide further insights into the evolutionary dynamics during faunal shifts, during extinctions and recoveries, as well as during the appearance of novel features. However, this also requires resolved phylogenetic hypotheses, which are lacking within and between many echinoderm groups. This

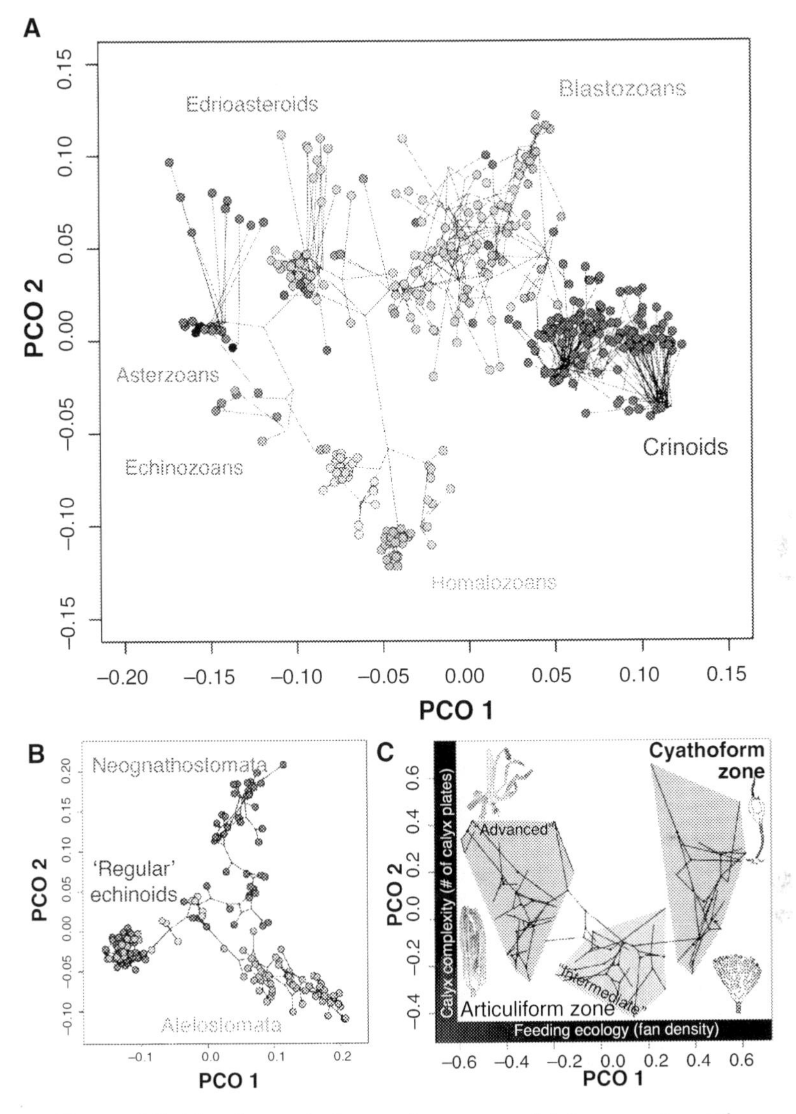


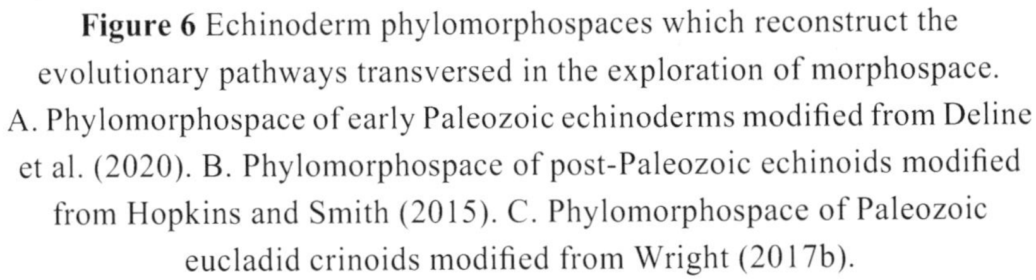


**Figure 6** Echinoderm phylomorphospaces which reconstruct the evolutionary pathways transversed in the exploration of morphospace. A. Phylomorphospace of early Paleozoic echinoderms modified from Deline et al. (2020). B. Phylomorphospace of post-Paleozoic echinoids modified from Hopkins and Smith (2015). C. Phylomorphospace of Paleozoic eucladid crinoids modified from Wright (2017b).

is often the result of difficulties in identifying homologies in disparate groups within echinoderms, thus the broad range of forms that make echinoderms morphologically enticing also hampers their study.

Exploring morphology within a phylogenetic prospective can also create a bridge to explore evolutionary change in reference to other data sets that utilize similar methods. For instance, faunal invasions and dispersal events have significant evolutionary impacts and have recently been examined using a phylogenetic prospective (Stigall et al., 2017; Stigall, 2019). Utilizing a multifaceted approach could allow the exploration of the rate of character change, changes in disparity, and directionality of change during different scales of faunal invasion or during different modes of speciation (dispersal or vicariance). In addition, as particular pathways of invasion are identified (Lam and Stigall, 2015), there is the opportunity to test the relationship between dispersal distance and degree of morphological change.

This prospective also allows the comparison of ecology and morphology in a very direct manner. Ecological disparity has been used to explore the relative complexity of organismal interactions through time (Novack-Gottshall, 2007). In many ways, the study of macroecology has been directly analogous to the study of morphology in that there has been a direct attempt to codify ecological habits through time. This has been used to build theoretical spaces and quantify their occupation through time, again mirroring studies of disparity (Bambach et al., 2007). Echinoderms provide an ample opportunity to compare the relationship between morphology and ecology in that they have been model organisms for studies of ecological interactions such as predation/parasitism (Baumiller and Gahn, 2002; 2004) and competition (Ausich, 1980; Ausich and Bottjer, 1982). Cole et al. (2019) used a phylogenetic approach to understand patterns of niche partitioning, competition, and ecospace occupation in an Ordovician crinoid community. Examining features associated with feeding ecology such as body size, filter size, and filter density provided insights into what features define ecological niches and which features are phylogenetically more divergent within closely related taxa (i.e. greater morphological plasticity). Combining this approach with patterns of morphological change would allow a comparison of evolutionary flexibility in form compared with ecological position and how this potentially shifts through time. In addition, it would allow the exploration of the morphological rate of change during large ecological events such as the origination of new predators (e.g. Sallan et al., 2011), the extinction of competitors (e.g. Ausich and Deline, 2012), or the shift into a novel ecological niche (e.g. Hopkins and Smith, 2015).

Overall, detailed morphological studies are an extraordinary and valuable tool in exploring macroevolutionary patterns and their underlying mechanisms, especially when paired with more expansive data sets (e.g. phylogenetic relationships, functional morphology, ecology, or biogeography). However, great care has to be taken in the experimental design and caution needs to be applied regarding the potential issues that can arise because of methodological choices.

# References

Atwood, J. W., & Sumrall, C. D. (2012). Morphometric investigation of the *Pentremites* fauna from the Glen Dean Formation, Kentucky. *Journal of Paleontology*, **86**(5), 813–828.

Ax, P. (1996). *Multicellular Animals: a new approach to the phylogenetic order in nature*, Berlin: Springer Press.

Ax, P. (2000). *Multicellular Animals: the phylogenetic system of the Metazoa*, Berlin: Springer Press.

Ax, P. (2003). *Multicellular Animals: order in nature-system made by man*. Berlin: Springer Press.

Ausich, W. I. (1980). A model for niche differentiation in Lower Mississipian crinoid communities. *Journal of Paleontology*, **54**(2), 273–288.

Ausich, W. I., & Bottjer, D. J. (1982). Tiering in suspension-feeding communities on soft substrata throughout the Phanerozoic. *Science*, **216**(4542), 173–174.

Ausich, W. I., & Deline, B. (2012). Macroevolutionary transition in crinoids following the Late Ordovician extinction event (Ordovician to Early Silurian). *Palaeogeography, Paleoclimatology, Palaeoecology*, **361–361**, 38–48.

Bambach, R. K., Bush, A. M., & Erwin, D. H. (2007). Autecology and the filling of ecospace: key metazoan radiations. *Palaeontology*, **50**(1), 1–22.

Bauer, J. E., Waters, J. A., & Sumrall, C. D. (2019). Redescription of *Macurdablastus* and redefinition of Eublastoidea as a clad of Blastoidea (Echinodermata). *Palaeontology*, **62**(6), 1003–1013.

Baumiller, T. K., & Gahn, F. J. (2002). Fossil record of parasitism on marine invertebrates with special emphasis on the platyceratid-crinoid interaction. *The Paleontological Society Papers*, **8**, 195–210.

Baumiller, T. K., & Gahn, F. J. (2004). Testing predator-driven evolution with Paleozoic crinoid arm regeneration. *Science*, **305**(5689), 1453–1455.

Brett, C. E., Moffat, H. A., & Taylor, W. L. (1997). Echinoderm taphonomy, taphofacies, and Lagerstätten. *The Paleontological Society Papers*, **3**, 147–190.

Briggs, D. E. G., Fortey, R. A., & Wills, M. A. (1992). Morphological Disparity in the Cambrian. *Science*, **256**(5064), 1670–1673.

Brower, J. C. (1974). Ontogeny of camerate crinoids. *University of Kansas Paleontological Contributions Papers*, **72**, 1–53.

Brusatte, S. L., Montanari, S., Yi, H. Y., & Norell, M. A. (2011). Phylogenetic corrections for morphological disparity analysis: new methodology and case studies. *Paleobiology*, **37**(1), 1–22.

Budd, G. E. (2006). On the origin and evolution of major morphological characters. *Biological Reviews*, **81**(4), 157–165.

Cailliez, F. (1983). The analytical solution of the additive constant problem. *Psychometrika*, **48**(2), 305–308.

Cherry, L. M., Case, S. M., Kunkel, J. G., Wyles, J. S., & Wilson, A. C. (1982). Body shape metrics and organismal evolution. *Evolution*, **36**(5), 914–933.

Ciampaglio, C. N. (2002). Determining the role that ecological and developmental constraints play in controlling disparity: examples from the crinoid and blastozoan fossil record. *Evolution and Development*, **4**(3), 170–188.

Ciampaglio. C. N., Kemp, M., & McShea, D. W. (2001). Detecting changes in morphospace occupation patterns in the fossil record: characterization and analysis of measures of disparity. *Paleobiology*, **27**(4), 695–715.

Cole, S. R. (2018). Phylogeny and evolutionary history od diplobathrid crinoids (Echinodermata). *Palaeontology*, **62**(3), 357–373.

Cole, S. R., Wright, D. F., & Ausich, W. I. (2019). Phylogenetic community paleoecology of one of the earliest complex crinoid faunas (Brechin Lagerstätte, Ordovician). *Palaeogeography, Palaeoclimatology, Palaeoecology*, **521**, 82–98.

Davidson, E. H., & Erwin, D. H. (2006). Gene regulation networks and the evolution of animal body plans. *Science*, **311**(5762), 796–800.

Deline, B. (2009). The effects of rarity and abundance distributions on measurements of local morphological disparity. *Paleobiology*, **35**(2), 175–189.

Deline, B. (2015). Quantifying morphological diversity in early Paleozoic echinoderms. In S. Zamora & I. Rabano, eds., *Progress in Echinoderm Palaeobiology*, Madrid: Instituto Geológico y Minero de España, pp. 45–48.

Deline, B., & Ausich, W. I. (2017). Character selection and the quantification of morphological disparity. *Paleobiology*, **43**(1), 68–84.

Deline, B., Ausich, W. I., & Brett, C. E. (2012). Comparing taxonomic and geographic scales in the morphologic disparity of Ordovician through Early Silurian Laurentian crinoids. *Paleobiology*, **38**(4), 538–553.

Deline, B., Greenwood, J. M., Clark, J. W., Puttick, M. N., Peterson, K. J., & Donoghue, P. C. J. (2018). Evolution of metazoan morphological disparity. *Proceedings of the National Academy of Sciences*, **115**(38), E8909–E8918.

Deline, B. & Thomka, J. R. (2017). The role of preservation on the quantification of morphology and patterns of disparity within Paleozoic echinoderms. *Journal of Paleontology*, **91**(4), 618–632.

Deline, B., Thompson, J. R., Smith, N. S., et al. (2020). Evolution and development at the origin of a phylum. *Current Biology*, **30**(9), 1672–1679.

Eble, G. J. (2000). Contrasting evolutionary flexibility in sister groups: disparity and diversity in Mesozoic Atelostomate echinoids. *Paleobiology*, **26**(1), 56–79.

Erkenback, E. M., & Thompson, J. R. (2019). Cell type phylogenetics informs the evolutionary origin of echinoderm larval skeletonogenic cell identity. *Nature Communications Biology*, **2**(1), 10–12.

Erwin, D. H. (2007). Disparity: morphological pattern and developmental context. *Palaeontology*, **50**(1), 57–73.

Ferrón, H. G., Greenwood, J. M., Deline, B., et al. (2020) Categorical versus geometric morphometric approaches to characterizing the evolution of morphological disparity in Osteostraci (Vertebrata, Stem-Gnathostomata). *Palaeontology*, **63**(5), 717–732.

Foote, M. (1991). Morphological and taxonomic diversity in clade's history: the blastoid record and stochastic simulations. *Contributions from the University of Michigan Museum of Paleontology*, **28**(6), 101–140.

Foote, M. (1992). Paleozoic record of morphological diversity in blastozoan echinoderms. *Proceedings of the National Academy of Sciences*, **89**(16), 7325–7329.

Foote, M. (1993). Contributions of individual taxa to overall morphological diversity. *Paleobiology*, **19**(4), 3013–419.

Foote, M. (1994a). Morphological disparity in Ordovician-Devonian crinoids and the early saturation of morphological space. *Paleobiology*, **20**(3), 320–344.

Foote, M. (1994b). Morphology of Ordovician-Devonian crinoids. *Contributions from the University of Michigan Museum of Paleontology*, **29**(1), 1–39.

Foote, M. (1995a). Morphological diversity of Paleozoic crinoids. *Paleobiology*, **21**(3) 273–299.

Foote, M. (1995b). Morphology of Carboniferous and Permian crinoids. *Contributions from the University of Michigan Museum of Paleontology*, **29**(7), 135–184.

Foote, M. (1996). Ecological controls on the evolutionary recovery of post-Paleozoic crinoids. *Science*, **274**(5292), 1492–1495.

Foote, M. (1997). The evolution of morphological diversity. *Annual Review of Ecology and Systematics*, **28**(1), 129–152.

Foote, M. (1999). Morphological diversity in the evolutionary radiation of Paleozoic and post-Paleozoic crinoids. *Paleobiology*, **25**(2) supplement, 1–115.

Gerber, S. (2019). Use and misuse of discrete character data for morphospace and disparity analysis. *Palaeontology*, **62**(2), 305–319.

Gerber, S., Eble, G. J., & Neige, P. (2011). Developmental aspects of morphological disparity dynamics: a simple analytical exploration. *Paleobiology*, **37**(2), 237–251.

Gould, S. J. (1989). *Wonderful Life: the Burgess Shale and the nature of history*, New York: WW Norton & Company.

Gould, S. J. (1991). The disparity of the Burgess Shale arthropod fauna and the limits of cladistics analysis: why we must strive to quantify morphospace. *Paleobiology*, **17**(4), 411–423.

Guillerme, T., & Cooper, N. (2018). Time for a rethink: time sub-sampling methods in disparity-through-time analyses. *Palaeontology*, **61**(4), 481–493.

Hetherington, A. J., Sherratt, E., Ruta, M., Wilkinson, M., Deline, B., & Donoghue, P. C. J. (2015). Do cladistic and morphometric data capture common patterns of morphological disparity? *Palaeontology*, **58**(3), 393–399.

Hopkins, M. J. (2017). How well does a part represent the while? A comparison of cranidial hape evolution with exoskeletal character evolution in the trilobite family Pterocephaliidae. *Palaentology*, **60**(3), 309–318.

Hopkins, M. J., & Gerber, S. (2017). Morphological disparity. In L. Nuño de la Rosa & G.B. Müller eds., *Evolutionary Developmental Biology*. New York: Springer International Publishing, pp. 1–12.

Hopkins, M. J., & Smith, A. B. (2015). Dynamics evolutionary change in post-Paleozoic echinoids and the importance of scale when interpreting changes in rates of evolution. *Proceedings of the National Academy of Sciences*, **112**(12), 3758–3763.

Hoyal Cuthill, J. F., & Hunter, A. W. (2020). Fullerene-like structures of Cretaceous crinoids reveal topologically limited skeletal possibilities. *Palaeontology*, **63**(3), 513–524.

Huelsenbeck, J. P., Nielsen, R., & Bollback, J. P. (2003). Stochastic mapping of morphological characters. *Systematic Biology*, **52**(2), 131–158.

Hughes, M., Gerber, S., & Wills, M. A. (2013). Clades reach highest morphological disparity early in their evolution. *Proceedings of the National Academy of Science*, **110**(3), 13875–13879.

Huttegger, S. M., & Mitteroecker, P. (2011). Invariance and meaningfulness in phenotype spaces. *Evolutionary Biology*, **38**(3), 335–351.

Jaanusson, V. (1981). Functional thresholds in evolutionary progress. *Lethaia*, **14**(3), 251–260.

Kammer, T. W., Sumrall, C. D., Zamora, S., Ausich, W. I., Deline, B. (2013). Oral region homologies in Paleozoic crinoids and other plesiomorphic pentaradial echinoderms. *PloS one*, **8**(11), e77989.

Lam, A. R., & Stigall, A. L. (2015). Pathways and mechanisms of Late Ordovician (Katian) faunal migrations of Laurentia and Baltica. *Estonian Journal of Earth Sciences*, **64**(1), 62–67.

Lefebvre, B., Eble, G. J., Navarro, N., & David, B. (2006). Diversification of atypical Paleozoic echinoderms: a quantitative survey of patterns of stylophoran disparity, diversity, and geography. *Paleobiology*, **32**(3), 483–510.

Lloyd, G. T. (2016). Estimating morphological diversity and tempo with discrete character-taxon matrices: implementation, challenges, progress, and future directions. *Biological Journal of the Linnean Society*. **118**, 131–151.

Lloyd, G. T. (2018). Journeys through discrete-character morphospace: synthesizing phylogeny, tempo, and disparity. *Palaeontology*, **61**(50), 637–646.

MacLeod, N. (2015). Use of landmark and outline morphometrics to investigate thecal form variation in crushed godiid echinoderms. *Palaeoworld*, **24**(4), 408–429.

Mitteroecker, P., & Huttegger, S. M. (2009). The concept of morphospaces in evolutionary and developmental biology: mathematics and metaphors. *Biological Theory*, **4**(1), 54–67.

Mongiardino Koch, N., Ceccarelli, F. S., Ojanguren-Affilastro, A. A., & Ramirez, M. J. (2017). Discrete and morphometric traits reveal contrasting patterns and processes in the macroevolutionary history of a clade of scorpions. *Journal of Evolutionary Biology*, **30**, 814–825.

Mooi, R., & David, B. (1997). Skeletal homologies of echinoderms. *The Paleontological Society Papers*, **3**, 305–335.

Novack-Gottshall, P. M. (2007). Using a theoretical ecospace to quantify the ecological diversity of Paleozoic and modern marine biotas. *Paleobiology*, **33**(2), 273–294.

Paul, C. R. C., & Smith, A. B. (1984). The early radiation and phylogeny of echinoderms. *Biological Reviews*, **59**(4), 443–481.

Raup, D. M. (1962). Computer as aid in describing form in gastropod shells. *Science*, **138**(3537), 150–152.

Raup, D. M. (1966). Geometric analysis of shell coiling: general problems. *Journal of Paleontology*, **40**(5), 1178–1190.

Raup, D. M. (1967). Geometric analysis of shell coiling: coiling in ammonoids. *Journal of Paleontology*, **41**(1), 43–65.

Raup, D. M., & Michelson, A. (1965). Theoretical morphology of the coiled shell. *Science*, **147**(3663), 1294–1295.

Riedl, R. (1977). A systems-analytical approach to macro-evolutionary phenomena. *The Quarterly Review of Biology*, **52**(4), 351–370.

Romano, M., Brocklehurst, N., Manni, R., & Nicosia, U. (2018). Multiphase morphospace saturation in cyrtocrinid crinoids. *Lethaia*, **51**, 538–546.

Runnegar, B. (1987). Rates and modes of evolution in the Mollusca. In K. S. W. Campbell & M. F. Day, eds., *Rates of evolution*. London: Allen and Unwin, pp.39–60.

Salazar-Ciudad, I. (2006). On the origins of morphological disparity and its diverse developmental bases. *Bioessays*, **28**(11), 1112–1122.

Sallan, L. C., Kammer, T. W., Ausich, W. I., & Cook, L. A. (2011). Persistent predator-prey dynamics revealed by mass extinction. *Proceedings of the National Academy of Sciences*, **108**(20), 8335–8338.

Schaeffer, J., Benton, M. J., Rayfield, E. J., & Stubbs, T. L. (2019). Morphological disparity in theropods jaws: comparing discrete characters and geometric morphometrics. *Palaeontology*, **63**(2), 283–299.

Sheffield, S. L., & Sumrall, C. D. (2019). The phylogeny of the Diploporita: a polyphyletic assemblage of blastozoan echinoderms. *Journal of Paleontology*, **93**(4), 740–752.

Sheffield, S. L., Zachos, L. G., & Lewis, R. D. (2012). A morphometric study of *Erisocrinus*(Crinoidea) using ArcGIS. *Geological Society of America Abstracts with Programs*, 44, 232.

Sidlauskas, B., (2008). Continuous and arrested morphological diversification in sister clades of characiform fishes: a phylomorphospace approach. *Evolution*, **62**(12), 3135–3156.

Smith, A. B., & Savill, J. J. (2001). *Bromidechinus*, a new Ordovician echinozoan (*Echinodermata*), and its bearing on the early history of echinoids. *Earth and Environmental Science Transactions of the Royal Society of Edinburgh*, **92**(2), 137–147.

Smith, A. J., Rosario, M. V., Eiting, T. P., & Dumont, E. R. (2014). Joined at the hip: linked characters and the problem of missing data in studies of disparity. *Evolution*, **68**(8), 2386–2400.

Sprinkle, J., & Collins, D. (2006). New eocrinoids from the Burgess Shale, southern British Columbia, Canada, and the Spence Shale, northern Utah, USA. *Canadian Journal of Earth Sciences*, **43**(3), 303–322.

Stigall, A. L. (2019). The invasion hierarchy: ecological and evolutionary consequences of invasions in the fossil record. *Annual Reviews of Ecology, Evolution, and Systematics*, **50**, 355–380.

Stigall, A. L., Bauer, J. E., Lam, A. R., & Wright, D. F. (2017). Biotic immigration events, speciation, and the accumulation of biodiversity in the fossil record. *Global and Planetary Change*. **148**, 242–257.

Sumrall, C. D., & Waters, J. A. (2012). Universal elemental homology in glyptocystitoids, hemicosmitoids, coronoids and blastoids: steps toward echinoderm phylogenetic reconstruction in derived blastozoa. *Journal of Paleontology*, **86**(6), 956–972.

Sumrall, C. D., & Wray, G. A. (2007). Ontogeny in the fossil record: diversification of body plans and the evolution of 'aberrant' symmetry in Paleozoic echinoderms. *Paleobiology*, **33**(1), 149–163.

Thompson, D'A. W. (1917). *On growth and form*, London: Cambridge University Press.

Thompson, D'A. W. (1942). *On growth and form*, London: Cambridge University Press.

Valentine, J. W. (1969). Patterns of taxonomic and ecological structure of the shelf benthos during Phanerozoic time. *Palaeontology*, **12**(4), 684–709.

Villier, L., & Eble. G. J. (2004). Assessing the robustness of disparity estimates: the impact of morphometric scheme, temporal scale, and taxonomic level in spatangoid echinoids. *Paleobiology*, **30**(4), 652–665.

Waters, J. A., Horowitz, A. S., & Macurda, D. B. Jr. (1985). Ontogeny and phylogeny of the Carboniferous blastoid. *Pentremites*. *Journal of Paleontology*, **59**(3), 701–712.

Webster, M., & Sheets, H. D. (2010). A practical introduction to landmark-based geometric morphometrics. *The Paleontology Society Papers*, **16**, 163–188.

Wills, M. A. (1998). Cambrian and recent disparity: the picture from priapulids. *Paleobiology*, **24**(2), 177–199.

Wright, D. F. (2017a). Bayesian estimation of fossil phylogenies and the evolution of early to middle Paleozoic crinoids. *Journal of Paleontology*, **91**(4), 799–814.

Wright, D. F. (2017b). Phenotypic innovation and adaptive constrains in the evolutionary radiation of Palaeozoic crinoids. *Scientific Reports*, **7**(1), 1–10.

Yochelson, E. L. (1978). An alternative approach to the interpretation of the phylogeny of ancient mollusks. *Malacologia*, **17**(2), 165–191.

Yochelson, E. L. (1979). Early radiation of Mollusca and mollusc-like groups. In M. R. House, ed., *The Origin of Major Invertebrate Groups*, Vol. 12, New York: Academic Press, pp.323–358.

Zachos, L. G., & Sprinkle, J. (2011). Computational model of growth and development in Paleozoic echinoids. In A. M. T. Elewa, ed., *Computational Paleontology*. Berlin: Springer, pp.75–93.

Zamora, S., Rahman, I. A., & Smith., A. B. (2012). Plated Cambrian bilaterians reveal the earliest stages of echinoderm evolution. *PLoS One*, **7**(6), e38296.

Zamora, S., Sumrall, C. D. & Vizcaïno, D. (2013). Morphology and ontogeny of the Cambrian edrioasteroid echinoderm *Cambraster cannati* from western Gondwana. *Acta Palaeontologica Polonica*, **58**(3), 545–559.

Zhao, Y., Sumrall, C. D., Parsley, R. L. & Peng., J. (2010). *Kalidiscus*, a new plesiomorphic edrioasteroid from the basal Middle Cambrian Kaili Biota of Guizhou Province, China. *Journal of Paleontology*, **84**(4), 668–680.

# Acknowledgements

I would like to thank W. Ausich, C. Sumrall, and S. Zamora for organizing the short course and inviting this review. My thoughts and views on echinoderm disparity and morphology have been shaped by many discussions with colleagues including but not limited to W. Ausich, C. Brett, D. Buick, P. Donoghue, M. Foote, D. Meyer, A. Miller, I. Rahman, C. Sumrall, J. Thomka, J. Thompson, and S. Zamora. Editorial assistance was provided by S. Deline. This manuscript was greatly improved with the helpful suggestions of H. Ferrón, I. Rahman, D. Wright, and S. Zamora.

Cambridge Elements

# Elements of Paleontology



## About the Series

The Elements of Paleontology series is a publishing collaboration between the Paleontological Society and Cambridge University Press. The series covers the full spectrum of topics in paleontology and paleobiology, and related topics in the Earth and life sciences of interest to students and researchers of paleontology.

**The Paleontological Society** is an international nonprofit organization devoted exclusively to the science of paleontology: invertebrate and vertebrate paleontology, micropaleontology, and paleobotany. The Society's mission is to advance the study of the fossil record through scientific research, education, and advocacy. Its vision is to be a leading global advocate for understanding life's history and evolution. The Society has several membership categories, including regular, amateur/avocational, student, and retired. Members, representing some 40 countries, include professional paleontologists, academicians, science editors, Earth science teachers, museum specialists, undergraduate and graduate students, postdoctoral scholars, and amateur/avocational paleontologists.

# Cambridge Elements

## Elements of Paleontology

### Elements in the Series

These Elements are contributions to the Paleontological Short Course on *Pedagogy and Technology in the Modern Paleontology Classroom* (organized by Phoebe Cohen, Rowan Lockwood and Lisa Boush), convened at the Geological Society of America Annual Meeting in November 2018 (Indianapolis, Indiana USA).

*Integrating Macrostrat and Rockd into Undergraduate Earth Science Teaching*
Pheobe A. Cohen, Rowan Lockwood, and Shanan Peters

*Student-Centered Teaching in Paleontology and Geoscience Classrooms*
Robyn Mieko Dahl

*Beyond Hands On: Incorporating Kinesthetic Learning in an Undergraduate Paleontology Class*
David W. Goldsmith

*Incorporating Research into Undergraduate Paleontology Courses: Or a Tale of 23,276 Mulinia*
Patricia H. Kelley

*Utilizing the Paleobiology Database to Provide Educational Opportunities for Undergraduates*
Rowan Lockwood, Pheobe A. Cohen, Mark D. Uhen, and Katherine Ryker

*Integrating Active Learning into Paleontology Classes*
Alison N. Olcott

*Dinosaurs: A Catalyst for Critical Thought*
Darrin Pagnac

*Confronting Prior Conceptions in Paleontology Courses*
Margaret M. Yacobucci

*The Neotoma Paleoecology Database: A Research Outreach Nexus*
Simon J. Goring, Russell Graham, Shane Oeffler, Amy Myrbo, James S. Oliver, Carol Ormond, and John W. Williams

*Equity, Culture, and Place in Teaching Paleontology: Student-Centered Pedagogy for Broadening Participation*
Christy C. Visaggi

*Understanding the Tripartite Approach to Bayesian Divergence Time Estimation*
Rachel C. M. Warnock and April M. Wright

*Computational Fluid Dynamics and its Applications in Echinoderm Palaeobiology*
Imran A. Rahman

A full series listing is available at: www.cambridge.org/EPLY

Printed in the United States
By Bookmasters